WILL WE SURVIVE CLIMATE CHANGE?
One Last Chance

Len Frenkel

Copyright © 2015 Len Frenkel.

All rights reserved. No part of this book may be reproduced, stored, or transmitted by any means—whether auditory, graphic, mechanical, or electronic—without written permission of both publisher and author, except in the case of brief excerpts used in critical articles and reviews. Unauthorized reproduction of any part of this work is illegal and is punishable by law.

ISBN: 978-1-4834-3345-5 (sc)
ISBN: 978-1-4834-3344-8 (e)

Library of Congress Control Number: 2015909781

Because of the dynamic nature of the Internet, any web addresses or links contained in this book may have changed since publication and may no longer be valid. The views expressed in this work are solely those of the author and do not necessarily reflect the views of the publisher, and the publisher hereby disclaims any responsibility for them.

Any people depicted in stock imagery provided by Thinkstock are models, and such images are being used for illustrative purposes only.
Certain stock imagery © Thinkstock.

Lulu Publishing Services rev. date: 06/23/2015

To all life on our one and only home, Earth.

Table of Contents

Keynote vii

Acknowledgments ix

Preface xi

Chapter 1   Introduction 1

Chapter 2   Climate Change 2

   What Is Climate Change? 3

   What Causes Climate Change? 4

   How Do We Know That Climate Change Exists and That It Is Caused by Human Activity? 6

   What Are Some of the Consequences of Climate Change? 9

      Flooding 9

      Storms 10

      Drought 11

      Adaptation 13

      Ocean Acidity 15

      Ocean Warming 17

      Species Extinctions 19

      Tipping Points 20

    Greenhouse Gases 21

    What Will Happen to Us? 24

Chapter 3 Livestock 26

Chapter 4 Where Are the Media?? 30

Chapter 5 Conclusion 34

Bibliography 38

Endnotes 40

Keynote

What have the media not told us about climate change? This book answers that question and makes a compelling, passionate, and informed plea for all of us to change one aspect of our lives in order to save ourselves and most life on the planet from the rapidly approaching effects of climate change.

Acknowledgments

My wife and partner, Ruth, has been with me all the way through this process and has been my editor in chief and cheering section. Lindsey Rorden, my wonderful granddaughter, has gone through the manuscript many times with a most critical eye and has made excellent suggestions for improvement. Thanks also to the following people for their critical and supportive suggestions: Jeff Anhang, Edith Heitmann-Tsacle, Sandra Lavini, Linda Marsa, Cara O'Sullivan, Leanne Oswald, Marissa Oswald, Elana Pessin, Vilma Reynoso, and Edward Silberman.

Preface

Dear Reader,

This book is about the reality of climate change. This reality, sometimes called climate crisis or climate chaos, is all around us, here in the United States as well as everywhere on the planet. It is not going away; in fact, it is getting worse by the day. We haven't experienced a worldwide crisis this serious since World War II when many millions of people faced bullets and bombs. This new war is against invisible and silent enemies—carbon dioxide and methane. It is absolutely necessary that you and everyone else in the world have some information about it so there can be the possibility of taking actions to stop the many climate threats to our planet. We are all in this together. I hope that you will be patient and read through this book, for there is a hopeful end to the story that I am about to share with you.

Thirty-four years ago, when I was single and living in South Jersey across the bay from Atlantic City, I went on a photography excursion in Brigantine National Wildlife Refuge. It was November but not terribly cold yet. I was alone, comfortably spending the day looking for subjects to photograph with my new close-up lenses. Across the parking lot, I happened to spot a group of people dressed in outdoor clothing, hats, and boots. I said to myself, "They look like my kind of people." Something stirred in me, so I went over to them and asked who they were and what they were doing there. They told me they were members of the Outdoor Club of South Jersey and about to start on a five- mile hike through the woods. Wow, did that appeal to me. Even though I had never done any hiking in my life, off I went with them—with the wrong shoes, no hiking pole, and no lunch. It was a daunting and thrilling experience and one that changed my life immensely. I continued to hike with the club, even getting into short backpack trips. Hiking and backpacking would be an important part of my life for the next thirty-plus years.

Not many years later, with my new wife who loved hiking and traveling as much as I did, I visited Glacier National Park in Montana for a few days of hiking. The glaciers were magnificent in their size. It was overwhelming to think about the amount of ice built up over tens of thousands of years. Just recently, we went back to see what changes occurred in the ensuing years, having read that climate change had reduced them in size. What a shock! The reports were correct. The glacial fronts had receded significantly, with the overall size and weight of ice greatly diminished. I had a feeling of tremendous sadness, paired with the realization that something dreadful was going on that was causing the glaciers to shrink.

A friend recently shared her experience with me. She said, "We were in Alaska in 2004, and all the rangers were teaching us about the change in the glaciers due to global warming and climate change. I think I remember they said by 2025 many glaciers the tourists visit will be so receded due to melting that you will not be able to see them anymore! So, so sad."

I had read reports of other glaciers around the world showing the same shrinking. After I returned home, I kept seeing images of these receding glaciers in various environmental magazines, which led me to research the subject to find the cause. I needed to know more. Much more.

On the left, a glacier not yet affected by warming. On the right, a glacier heavily receded by melting.

I. Introduction

The first part of this book discusses causes and consequences of climate change, followed by information about livestock production and its very serious role in climate change. The next chapter focuses on possible reasons why we haven't been informed by the major media about these issues. The conclusion explains how each of us can help avoid the worst effects of climate change by taking one simple action.

This book is not meant to be a thorough description of any of these three topics, but it is meant to be an exploration into these subjects so that we are all informed citizens capable of making good decisions for the present and especially for the future.

You may be wondering why crucial information about climate change hasn't gone public. Why don't the major news media outlets (TV, radio, magazines, and newspapers) inform us about this serious problem? Why don't we hear and read about it every day if it's so threatening? At the end of this book is a chapter answering these very questions. In part, this is why I decided to write this book: to inform you and millions of others about this subject because you haven't been informed to the extent that you need and deserve.

Recently, I edited a book written by a friend on these three subjects[1]. His book is written in textbook format, meant for professionals and students. I wanted to write a book for the general public, using a comfortable style and everyday language. No dictionary or graduate degree required. This is it.

II. Climate Change

You've probably noticed that weather patterns have been changing in recent years, most pronounced in the last decade. More frequent and heavier storms, and more damage from them, are taking place around the world. Spring starts earlier, and fall starts later. The growing season is longer than it used to be. Maybe you've seen more insects than in the past. You may have noticed ticks and roses in December in the eastern United States. Who would have thought? Do you know about the drought conditions in our Southwest? Los Angeles, I read, had no rain in January 2014, generally the rainiest month of the year.[2] Many, many days, I see in the weather section of my local newspaper that Anchorage, Alaska, has warmer winter weather than Boston, New York City, and Chicago. How could this happen? And, of course, there are the glaciers. They exist not just in the United States but all around the world, in the Andes Mountains of South America, the Alps in Europe, and the Himalayas in Asia. They are all rapidly melting. The loss of glacial ice and changing weather patterns are not a coincidence. They are results of the atmosphere getting warmer. Look at how Alaska's average temperature has changed over the last sixty-five years, the image supplied by the Alaska Climate Research Center.

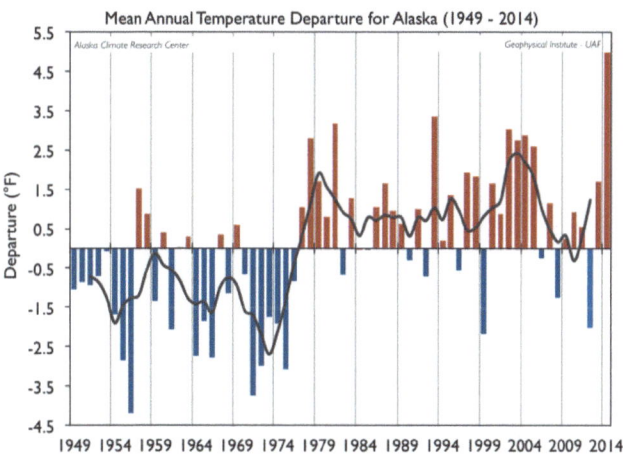

A. What Is Climate Change?

It is important to differentiate between weather and climate. So many people have difficulty with this concept. Weather is what is happening in your community, county, or state. It is short-term: what is happening today or this week. Are we having sunshine, rain, snow, heat, or cold these next few days? Climate applies to larger regions and is longer-term. Is the southwest United States going to continue to have severe drought? Will the Northeast continue to have lots of rain and snow in the next years? Will Australia continue its drought and Central Asia lose its glaciers? Will the northern polar regions continue to warm faster than the temperate regions? These are climate questions.

Now for an explanation of climate change. Heat, light, and ultraviolet rays from the sun strike the earth constantly. Much of the heat is radiated back into space, but certain gases in the atmosphere prevent some of it from leaving earth's atmosphere. This process acts like a greenhouse or a plant nursery, where plants can be grown in a warm climate even in the coldest days of winter. The illustration provides a simplified version of a very complex interaction of radiations and gas particles.

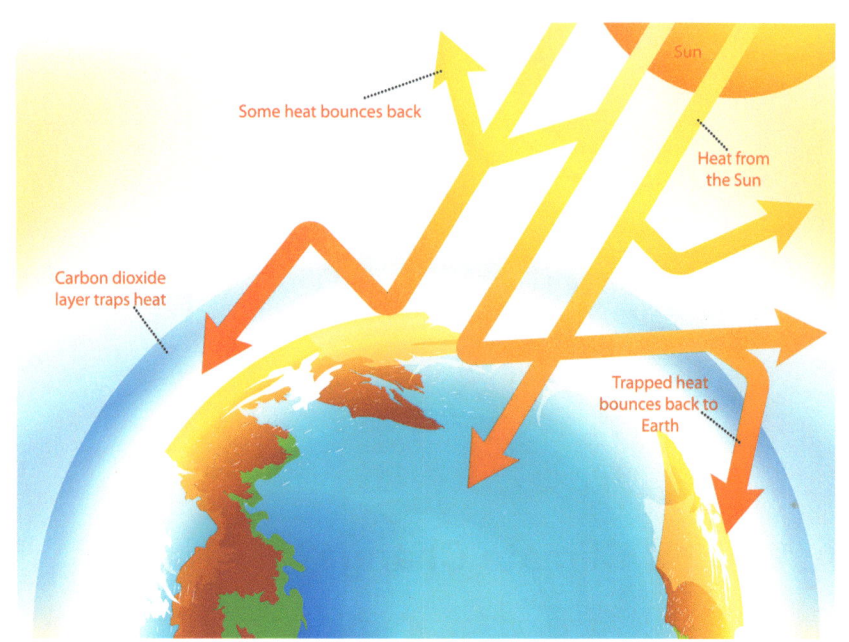

The balance between gain of heat and loss of heat allows life to exist on our planet. Anything that interferes with this balance disrupts the atmosphere and consequently life itself.

B. What Causes Climate Change?

Since the beginning of the Industrial Age, about two hundred years ago when coal and oil became important sources of heat and energy for industry, these fossil fuels have been burned in enormous quantities, producing carbon dioxide. Carbon dioxide is a colorless, odorless gas that is poisonous to life at high levels. But it is also what the leaves of plants absorb in order to manufacture additional plant material and produce the oxygen that we need for life: photosynthesis. A certain low level of carbon dioxide has been in the atmosphere since the end of the last Ice Age, and it is the primary gas that keeps our planet warm. That is why carbon dioxide is called a greenhouse gas; it helps keep the planet warm and comfortable, on average. Without it, we would have an ice age. With too much, we would have a roasting age. More about carbon dioxide in a later section.

The problem is that we now have too much carbon dioxide in the atmosphere. This is a direct result of burning tremendous amounts of coal and oil to provide us with a good life with plenty of food, loads of products, travel, and comfortable places to live. The excess carbon dioxide that has been building up has now been warming our atmosphere above a safe level. Scientists who study this subject have many ways of measuring the actual warming as well as determining the level of carbon dioxide in the atmosphere.

Another gas that acts as a greenhouse gas is methane, and it is also colorless, odorless, and poisonous. It is the main part of natural gas that we use for heating homes and schools, making glass, and many other manufacturing processes. A tremendous amount of methane has been trapped in the ground in Arctic regions, but as the earth warms, methane is being released into the atmosphere. It is a much stronger greenhouse gas than carbon dioxide. Now here's a funny source of methane: when cows digest their food, they expel methane from both ends of their digestive system. Given that there are millions and millions of cows around the world at any one time, that's a lot of methane being released. So now we have a second, important contributor to climate change. More information about methane in a later chapter.

C. How Do We Know That Climate Change Exists and That It Is Caused by Human Activity?

Recently I was watching a PBS program, and an author by the name of Mark Lynas was being interviewed. He had written a book called *Six Degrees: Our Future on a Hotter Planet*.[3]

I was so disturbed by what he had to say about what is occurring and is likely to occur in the future on our planet that I had to read his book. In it, there are six chapters, each devoted to the earth's atmosphere warming by one degree centigrade. Each one-degree change in centigrade corresponds to a 1.8-degree change in Fahrenheit. Scientists use the centigrade scale, so it requires some getting used to. For example, if the average global temperature rises by two degrees centigrade (C), that corresponds to an increase of about 3.6 degrees Fahrenheit (F). Going back to the book, each chapter showed increasing environmental problems from the warming. I managed to read four chapters, which was more than enough to be extremely worried about our future. This was the point at which Lynas indicated that life on the planet was pretty much over, according to his evaluations of the scientific research that had been published to that date. He described a world in which cities such as Alexandria (Egypt), Boston, New York City, Shanghai, Mumbai (India), London, and Venice will likely be flooded. The West and East Antarctic Ice Sheets will be well on the way to melting sufficiently to flood even more regions worldwide that will be beyond human ability to overcome. More food will be needed for increasing populations at the same time that food crops such as rice, wheat, and corn will be reduced due to the warming. Desertification will expand, leading to starvation and water losses.

That information and so much more was the trigger for me to become active in the climate change movement and eventually to write this book. There are thousands of scientists from universities, government research agencies, and nongovernment organizations that study climate change. They are from all over the globe—Europe, Asia, North and South America, Africa, Australia. There are oceanographers, physicists, chemists, hydrologists, geologists, computer scientists, climatologists, astronomers, biologists, and paleontologists who carry out investigations and report their findings. One of the most important groups that collect information from all these investigators is the Intergovernmental Panel on Climate Change (IPCC), sponsored by the United Nations. The IPCC meets every five years to evaluate the latest data and issue a report on its findings. Since the IPCC's formation in 1988, these reports have shown increasing confidence that climate change is real and is caused by human activities.

Scientists have found many times over and from many different sources that the increasing temperature of the atmosphere strongly corresponds to the increased burning of fossil fuels over the last two hundred years. The last fifty to seventy-five years have shown a significantly greater rate of warming. The graph below shows increasing carbon dioxide (black curve) corresponding to planetary warming (red region). The graph comes from a National Climatic Data Center, National Oceanic & Atmospheric Administration (NCDC) (NOAA) report.

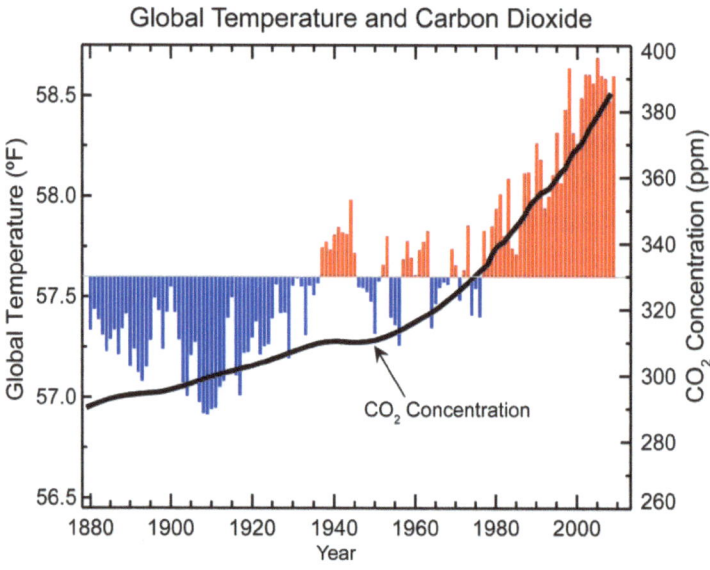

Oceanographers have found that the oceans also have warmed due to increased absorption of heat from the atmosphere. This may account for the leveling of the temperature in the lower atmosphere time period. (It has very recently been found to be due to errors in measuring atmospheric temperatures. Corrections show that temperatures continued to rise.) Biologists, paleontologists, geologists, and climatologists have found fossils in rock layers under the oceans as well as deep in ice that they can analyze for their carbon content to learn what species lived during what time period in order to determine the climate at that time. They know that certain species like warm weather while others prefer cold. Ice analysis shows the amount of trapped carbon dioxide and its age in ice recovered from great depths in Arctic and Antarctic regions. Satellites containing equipment measuring recent carbon dioxide and methane levels have found increasing temperatures and gas levels. These and so many more investigations come together to conclude that the atmosphere is warming faster than in the last 100 million years of the earth's history, except for when a comet slammed into the Yucatan peninsula in Mexico some 65 million years ago. The comet was responsible for sudden, large-scale atmospheric heating, which led to the extinction of many species, most notably the dinosaurs. But aside from that kind of sudden climate change, today's science shows that our atmosphere hasn't been this warm in at least the last 800,000 years.[4]

D. What Are Some of the Consequences of Climate Change?

1. Flooding

Unfortunately, there are so many consequences of the warming of the atmosphere that it is difficult to decide where to start. Let's go back to the glaciers. All glaciers are melting faster than new ice is forming. We know that they all slide down the mountains until the front edge drops into a river, bay, or ocean. They've been losing ice faster and faster as the atmosphere and oceans get warmer. The short-term effect is that they are causing ocean levels to rise, which means that low-lying islands and ocean communities will be swamped or flooded in coming years. Many coastline cities around the world are threatened by flooding. This picture shows what we can expect to be a common experience in the near future.

Glacial water is needed by billions of people around the world as their source of fresh water for drinking, growing food crops, and various manufacturing operations. As the glaciers recede, there will be less and less fresh water available. After the rapid melting of the glaciers and the simultaneous flooding, many rivers will have less flow and will possibly even dry up during increasingly hotter weather.

2. Storms

We know from scientific reports as well as from personal experience that we in the United States, as well as in other locations, are experiencing storms that take place more frequently and that are stronger than in the past. This has resulted in much damage. These storms are expected to get worse over the years as the planet's warming continues. When a storm is particularly severe, flooding results, which can wash away people, homes, schools, agricultural fields, and businesses. As reported by the television news channel CNN, in the 2000 decade, there were eleven major storms worldwide with winds over 125 mph.[5]

The Union of Concerned Scientists reported an increase of 74 percent in the amount of rain or snow that fell in the United States' Northeast in the last fifty years.[6]

There has also been an increase in the strength and frequency of snow storms, particularly in the northeastern United States. As I write this in February 2015, Boston has had an unprecedented amount of snow falling this winter, closing down the city for periods of time and wreaking havoc on all forms of transportation. There is little doubt that these disruptions in weather patterns over large areas are a result of anything other than climate change. This pattern is taking place around the world as well.

3. Drought

Unexpectedly, areas of drought are increasing at the same time as flooding takes place in other locations. These are all predicted by models that scientists use to understand climate change. The Amazon rain forest, for example, supplies moisture into the atmosphere that comes down as rain and snow when it heads west over the Andes Mountains. As the Amazon dries, as is now beginning to happen due to atmospheric warming as well as deforestation, this situation is getting worse. Less rain and snow will fall on the Andes Mountains, which is the primary source of fresh water for Peruvians living on the west coast of South America. Some researchers predict that in a best-case scenario, at least 20 percent of the Amazon rain forest will turn to savannah by century's end, while as high as 85 percent could be gone in a worst-case scenario.[7] The following photo shows a region in the Amazon that is normally water-filled.

Drylands, such as in our American southwest, as well as in desert areas of Africa, Australia, and parts of Asia, are becoming drier. Here's a picture of a farmer facing drought on his farmland.

Desertification is the process of dry lands becoming drier from many causes, especially climate change. Half the planet's land surface is considered dryland, with one-third of the human population living in these areas. Desertification is expected to continue and get worse. As examples, China has lost over 36,000 square miles to desertification while Africa lost 1374 square miles each year from the mid-1990s to 2000.[8] Droughts have worsened in Australia, southern Africa, the Sahel region of Africa, south Asia, and the Mediterranean region. Economic and personal costs can be enormous, such as loss of farmland, water supplies, and living space. The island of Taiwan is currently experiencing a record drought with the lowest rainfall in seventy years.

4. Adaptation

Almost all animal species have a limited tolerance for higher temperatures in their environment. With a warming climate everywhere, they try to avoid the heating by expanding their habitat to a cooler region. Many species are unable to make this change. They may not be able to find food or reproduce and thus go extinct. Some, for example, live on mountains, where there may be a limit as to how high they can live before there is no more space. As an example, pikas, cute little rodents living in mountainous areas of the American west, cannot survive temperatures above seventy-five degrees F. The warming from climate change drives them up the mountain where they may run out of living space. They are presently threatened as a species. The nine-banded armadillo, previously a strictly southern American species, is now found in North Carolina. Marmots are coming out of hibernation three weeks earlier than they did thirty years ago, and Canadian red squirrels are breeding eighteen days earlier, according to several reports. These are all the effects of a worldwide warming climate.[9] They may seem like unimportant examples, but one must keep in mind the interdependence of all species. Changes in the behavior of one species are very likely to change the living situation for other species.

University of Texas researchers Camille Parmesan and Gary Yohe made detailed calculations of the movements of 1700 species and found that, on average, a shift toward the cooler poles came to four miles per decade, and retreat up mountains came to twenty feet per decade. They also determined that spring was arriving 2.3 days per decade earlier.[10] These are unprecedented habitat movements.

Though some plants have shown better growth in a warming climate, many do not. Ultimately, as temperatures rise rapidly, almost all will not be able to photosynthesize sufficiently to survive.

In addition to the movements of land animals, disease-carrying insects are beginning to expand north into areas where they have never before been common. Mosquitos, ants, and wasps are some of these creatures. In the future, people and animals in temperate zones are likely to see malaria and other tropical diseases emerge. Plants are also having difficulty adapting to climate change. Pests on trees have decimated forests. For example, the mountain pine beetle carries bacteria lethal to lodge pole pine and ponderosa pine in the American northwest and into Canada. Millions of acres of forest have been destroyed as a result, putting pressure on the lumber industry that supplies wood for construction. The brown areas in the picture show trees destroyed by the beetles. The image is from the California Oak Mortality Task Force.

Diseases such as sudden oak death and needle blight in British

Columbia are being observed in areas where they were previously not present. The yellow star thistle, a California weed, has expanded its territory northward. Entire ecosystems, then, are destroyed or damaged, affecting many species of animals and plants living in the area. Other plants and animals that depend upon these affected regions for food, protection, and reproduction cannot survive.[11] The rate of warming, unprecedented in human history, makes adaptation very difficult for many species. This pattern will continue in the future.

5. Ocean Acidity

Most of the carbon dioxide released into the atmosphere is absorbed by the oceans. Let me show you the devastating effects that this has on ocean life. First, a little chemistry. A system of measuring acidity or alkalinity of a liquid is known as the pH scale. Completely neutral is pH 7.0, which is exactly in the middle of the scale running from zero to fourteen. The lower the number, the more acidic. The higher the number, the more alkaline. Each unit change means a ten times change in acidity or alkalinity. So a liquid with a pH of 6.0 is ten times as acidic as one that is neutral at pH 7.0. Some examples: stomach acid, lemon juice, and vinegar have a pH of around two; coffee comes in at around five; human blood 7.4 and household ammonia around eleven. Now here's the point of all this: before industrial times, ocean water had a pH of 8.2; today it is on average 8.1. This doesn't sound like much, but it means that it is 26 percent more acidic than it was before industry began releasing carbon dioxide in such enormous amounts. I'll get into those amounts very soon; get ready to be shocked by the numbers.

So what's the big deal about a little more acidity in the ocean? Many ocean creatures are very sensitive to the pH of the water in which they live. Many shelled organisms such as clams, oysters, and coral are having difficulty building their calcium carbonate shells in that more acidic water. In fact, some shelled creatures have had their shells dissolved by the water, which, of course, kills them. Because this acidification has taken place so quickly, these animals have not had time to adapt to the new conditions and thus are severely threatened. Clams and oysters constitute major food supplies and form the basis for large industries. Pteropods, teeny, shelled animals, are a major food supply for fish and other sea creatures. If they decrease in numbers, being at the bottom of the ocean food chain, then organisms higher on the food chain will perish. One prominent researcher predicts that ocean pH levels will drop to 7.8 by the end of this century, virtually guaranteeing the death in large numbers of most ocean species, including animals such as sharks, dolphins, whales, tuna, and salmon.[12] It is expected that perhaps millions of species will go extinct as a result.

The Great Barrier Reef, off the east coast of Australia, is the best example of an ocean structure based upon corals, which are acid-sensitive animals. When they die in significant numbers, the results are devastating. The photos show a live coral ecosystem and then a dead one next to it that has been acidified.

Here's a list of the animals dependent upon a living coral ecosystem: six hundred types of coral, three thousand kinds of shelled creatures, over 1,600 fish species, more than 130 shark species, and over thirty species of whales and dolphins.[13]

Continued release of large amounts of greenhouse gases will make these conditions worse.

6. Ocean Warming

Another result of climate change is warming of the oceans. Three aspects of ocean warming are important to discuss. First is the increasing evaporation of sea water as a result of a rise of ocean temperatures. Second is the effect ocean warming has on sea life, and the third is the rising of ocean levels.

How much warming is taking place? More than 90 percent of solar heat is absorbed by the oceans.[14] On average, this has caused only an increase of 0.2 degrees F/decade increase close to the surface.[15] Although this increase doesn't seem like much, it causes more water to evaporate into the air, leading to more rainfall, more snow, more storms, and disruption of wind circulations. Hurricanes and typhoons can cause serious destruction when they hit land, leading to deaths and losses of farms, homes, schools, and businesses. Some researchers found that North Atlantic waters had gone up in temperature about 2.3 degrees F. since records were first kept, which signals more evaporation, more storms, and more damage. These temperature increases also lead to faster melting of glaciers as well as the Arctic and Antarctic ice caps and the Greenland ice sheet.

Look at what has happened to the Greenland ice sheet in just the last ten years. Thanks to Thompson-Reuters News for this pair of images.

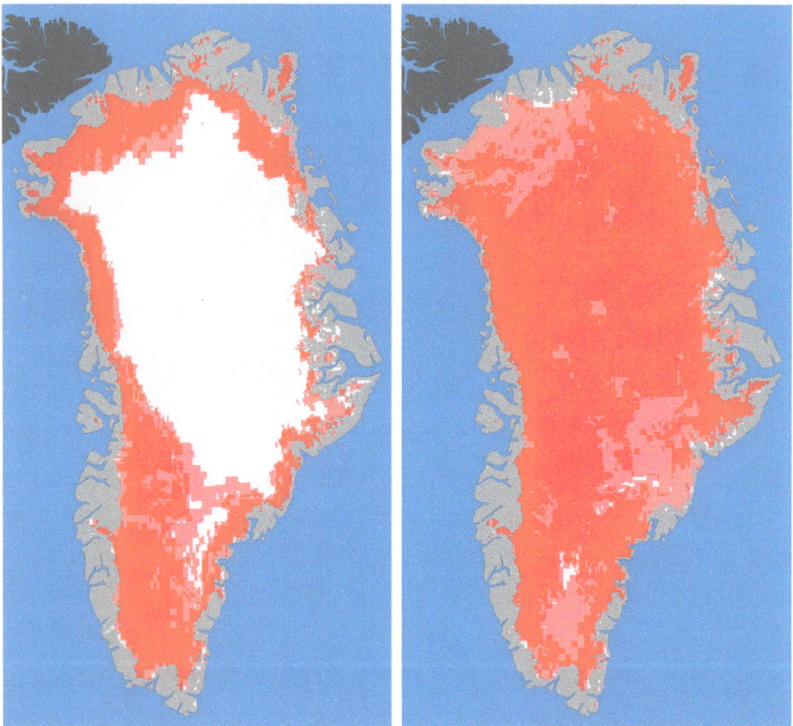

There have been massive losses of seasonal ice in Arctic waters, with forecasts that they could be completely gone in twenty years. In the past, there was ice even in the summers, but that situation is changing rapidly as the oceans warm. Many animals and fish depend upon this ice for their sustenance, including polar bears, seals, penguins, and walrus, now all threatened with extinction.

How much has the ocean risen since the Industrial Age kicked in? Core samples have told oceanographers that oceans have risen between four and eight inches in the last hundred years (.04 to .08 inches/year on average), but the rate of ocean rise has increased dramatically to 0.13 inches per year since the 1990s.

A recent study concluded that ocean levels will rise between 2.5 and 6.5 feet by the end of the century, enough to flood many coastal cities in the United States and abroad.[16] No countries are currently prepared for such a phenomenon, and none are likely to be able to defend against it.

The lives of many ocean animal species are very sensitive to temperature changes. Some are able to move to cooler water, but others cannot. Most corals, for example, are unable to tolerate temperatures over seventy-five degrees Fahrenheit, so when it gets too warm for them, they die. Because corals are not able to shift to cooler waters, all the other organisms that depend upon them for food and shelter perish along with them. Other ocean-living fish threatened by warming waters include salmon, anchovies, sardines, and mackerel. These fish represent very important food sources for humans and also serve as food for other species such as tuna, swordfish, sharks, and whales. Another creature, krill, a small shrimplike animal, doesn't do well in warming waters and must also move to cooler waters. They are the primary food for several whales, among them blue, humpback, and right whales. Krill are threatened by rising water temperatures. One group of researchers, Boris Worm of the University of Halifax with colleagues in the United States, the United Kingdom, Panama, and Sweden, has projected that by 2050 the tropical waters of the world may be completely empty of fish.[17]

7. Species Extinctions

Examples of potential species extinctions have already been shown in the section about adaptations, but because of the seriousness of this climate change consequence, more information is necessary. There have always been extinctions of species, largely when catastrophic events took place over the eons of the existence of life on earth. In the absence of these global catastrophes, scientists have determined that the normal extinction rate or background rate is between one and five species each year during the last millennia. However, since climate change kicked in, the situation has changed dramatically. Now, the rate of extinction is anywhere from one thousand to ten thousand times the background rate. Dozens of species disappear each day. Between 30 percent and 50 percent of known species are anticipated to be gone by midcentury. Gone forever, never to return. Worst of all, 99 percent of threatened species—that is, those that are expected to go extinct in the very near future—are dying out due to human activity.

Amphibians are being particularly hard hit, dying off at 25,000 to 45,000 times the background rate. Twelve percent of birds worldwide are threatened with extinction. Fish don't escape the downward slope, as 39 percent of freshwater fish are at risk, and globally 21 percent are in danger of extinction. Half of mammal species are in decline while one-fifth are at risk of disappearing forever. Of known reptiles, 21 percent are endangered or likely to go extinct. In the plant kingdom, consisting of greater than 300,000 species, 13,000 plants were evaluated as to their possibility of going extinct, and the researchers found a 68 percent likelihood.[18]

Because of the interdependence of plant and animal species, these forecasts indicate catastrophic consequences.

8. Tipping Points[19]

A tipping point is a step or stage in a process at which the process becomes irreversible, and a new situation becomes the normal. It can be thought of as a turning point. A tipping point can lead to a better situation, or it can lead to a new, negative situation. When applied to climate change, it is generally leading to a worsening of the climate. The one positive tipping point that I can think of is the one that I propose in the conclusions section.

There have been many tipping points discovered in the science of climate change, but several stand out as being particularly critical. Dr. James Hansen, recently retired from NASA and one of the most outspoken scientists warning us of the threat of climate change, thought that when carbon dioxide in the atmosphere reached 396 ppm in May 2012, carbon dioxide in the atmosphere had reached a tipping point.[20] By that, he meant that no matter what efforts are made to halt climate change, it is irreversible. At some point in the melting of the Greenland ice sheet and the West Antarctic Ice Sheet, no matter what we do, they will continue to melt and cause oceans to rise as well as many other terrible consequences. At some time during the dieback of the Amazon rain forest, a time will be reached when it will be impossible for the forest to be saved. The consequences of that will be almost unimaginable. The loss of permafrost in Siberia that releases methane gas will, at some time, be irreversible, triggering uncontrolled methane release. The slowing of the Gulf Stream northward along the East Coast of the United States has been observed. Continued slowing will have terrible results, so there will be a tipping point beyond which the slow current will not be reversible. Finally, the warming and acidification of the oceans will reach a level beyond which nothing can be done to stop those processes.

One of the bad things about these tipping points, and there are many, is that any one of them tends to lead to others. The result is likely to be the end of life on the planet.

E. Greenhouse Gases

The two most important greenhouse gases are carbon dioxide (CO_2) and methane (CH_4). They are called greenhouse gases because of their ability to reflect radiated heat from the earth's atmosphere back to the surface, keeping us warm, even considering that we have lots of cold and hot regions. The angle of the earth in its orbit around the sun contributes to the variations in temperature from season to season, as does the latitude of any particular location, but these are not factors in climate change. The temperature on earth gets warmer closer to the equator and colder farther from the equator. Between 1970 and 1999, the average surface temperature rose between .06 and .11 degrees C. per decade. This represents a very rapid rise in temperature in a short period of time.[21] There isn't much reason to believe that this pattern of rapidly rising temperature will not continue.

Why is this happening? Very simply, temperatures are rising because we are burning increasing amounts of fossil fuels (coal, oil, and natural gas), which produce more carbon dioxide than can be absorbed by trees. Methane levels are also increasing, due to the warming of deep Arctic Ocean sources. The gas is now also increasingly being released from the arctic tundra where it formed by the non-oxygen decay of plants over hundreds of millions of years. Large increases in methane levels also come from the vast numbers of cows raised for food. More about that in the next chapter.

Before the Industrial Revolution began two hundred years ago, very little excess carbon dioxide was formed from human actions. The highest level of carbon dioxide in the atmosphere was around 280 parts per million (ppm) and the lowest around 180 ppm, but it has now reached 400 ppm.[22] It is rising approximately 2 ppm each year. The increase from pre-industrial levels is almost a 50 percent increase and largely accounts for our present climate chaos. The 400 ppm level has been found to be the highest in the last 400,000 years and possibly the highest in the last 20 million years.[23] There is agreement among most researchers that the safest level of CO_2 would be 300 ppm. Higher levels continue to threaten much life on our planet. The following chart shows how carbon dioxide release into the atmosphere has increased over the years 1980 to 2009, a mere twenty-nine years, from about 1500 million tons/year to almost 8000 million tons/year.

As you might predict, it will continue to increase at this rapid rate. Image source is the Carbon Dioxide Information Analysis Center (CDIAC).

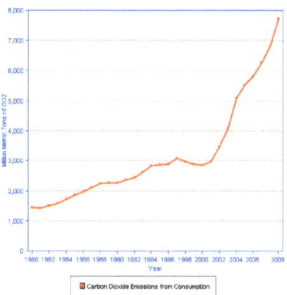

Although methane is much less common in our atmosphere than carbon dioxide, it is twenty to one hundred times as powerful a greenhouse gas. Methane doesn't last very long in the atmosphere, oxidizing to carbon dioxide fairly rapidly, but while it's there, it does its dirty work of keeping the sun's heat in our atmosphere. Over the last 400,000 years, the level has ranged from about 400 parts per billion (ppb) to highs of around 700 ppb.[24]

Recent measurements put it at around 1800 ppb, a very large increase, again, since the start of the Industrial Revolution. Many researchers now propose that methane's contribution to climate change is now greater than that of carbon dioxide.[25, 26] Almost all media discussions of climate change, as few as they are, focus on carbon dioxide. The graph that follows shows how methane has dramatically increased in the last fifteen years. Graph courtesy of the New Zealand Ministry of the Environment.

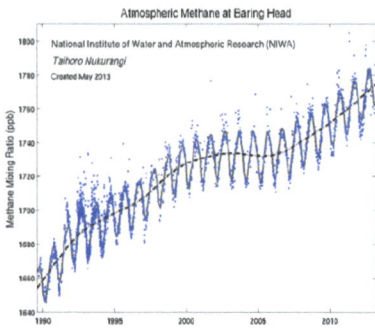

We tend not to be aware of the gases in our atmosphere, so some information about them might be interesting. The total weight of our atmosphere is 6,000,000,000,000,000 tons (quadrillion), of which carbon dioxide is currently around 3,000,000,000,000 tons (trillion). In 2010, calculations showed that 46,000,000,000 tons (billion) of greenhouse gases were released into the atmosphere. It is very likely that this figure is above 50 billion tons as of this writing. This greenhouse gas release is far in excess of the balance that has existed for thousands of years and which has caused the problems discussed in the consequences section, above.[27]

F. What Will Happen to Us?[28, 29]

If we do nothing to slow or stop the atmospheric warming all around our planet, we can expect by the second half of this century some devastating changes in living conditions. The most common and important grain foods—corn, rice, wheat, millet, and sorghum—will be less able to withstand the warming and will go into lower production. Other non-grain food plants will also have difficulty growing in a warmer climate and will become less available.

There will be more water in the warmer atmosphere as well as fewer glaciers that supply water to major rivers. Thus, there will be less water available for agriculture during the growing season.

With diminished river water, less will be available for household use as well as for transportation.

Rising ocean levels will flood shore communities around the world, causing loss of life, structures, and farmland. Simultaneously, people will be attempting to move away from these areas, putting pressure on inland communities to find products and services for them.

While the oceans are rising, they are also becoming warmer and acidic. The loss of marine life will be severe, causing food problems for those people dependent upon those resources.

Diseases and parasites now confined to tropical areas will expand into semitropical and temperate regions. People are not likely to have resistance to them, nor are medical communities likely to be prepared for them.

Drought regions will expand, forcing people to move away in order to find livable areas.

With grains and soybeans in limited supply, people will have to choose between feeding themselves and feeding their livestock, which now consume the majority of our grain production. Meat supplies will dwindle.

All in all, life will become very difficult. Some people who study and think about this future situation predict wars over resources: oil, food, water, and living space. [30, 31]

III. Livestock

Let's first define what we mean by livestock. Livestock are the animals raised for food, in the United States primarily cows, pigs, chickens, and turkeys. In many other countries, people and industries also raise rabbits, camels, sheep, goats, and buffalo for food. Chickens, pigs, and cows are the primary contributors in the United States to the greenhouse gases that cause climate change.

The quantities of animals raised for food are enormous, totaling over 10 billion land-based animals in the United States every single year. Of this amount, 10 billion are chickens, 40 million are cattle, and 134 million are pigs as of 2010. Worldwide, about 65 billion animals are killed for food each year.[32] To feed these animals, in 2000, 38 percent of the world's grains, 70 percent of the United States' grains, and 80 percent of the United States' corn were fed to livestock.[33] Livestock products are valued at over $100 billion in the United States each year.[34]

There is considerable controversy surrounding the degree to which livestock production affects climate change, but there is no doubt that it is very significant. It is an issue that is almost totally ignored by researchers, reporters, and even the most important and influential environmental organizations. In the last couple of years, several mainstream newspaper and magazine articles have been published in which the authors have pointed out the connection between livestock production and climate change. However, this issue is still a long way from getting the attention needed for a subject of such universal importance.

Raising animals for food is detrimental to not only climate stability but also to soils and waterways. Our emphasis in this book, however, is on the detrimental effects on the atmosphere.

Let's look at the larger picture regarding the greenhouse gases released as a result of livestock production. Total annual greenhouse gas production is agreed by most to be about 46 billion tons. The Food and Agriculture Organization of the United Nations (FAO) has determined that livestock production contributes 14.5 percent to total greenhouse gas levels (7.5 billion tons).[35] Earlier it had reported an 18 percent contribution,[36] while researchers at the Worldwatch Institute have calculated that it comes to at least 51 percent (32.5 billion tons).[37] It seems that different data have been used by each group of researchers to determine their percentages. All three percentages indicate that this is a very serious source of greenhouse gases. This information is generally not provided to the public.

Following is a list of some of the steps involved in producing animals for food. Researchers in this area debate which of these should be considered in determining the activities to be included in figuring out the importance of animal production to climate change.

- Gasoline is needed to power tractors to till soils to grow the main animal foods, corn, alfalfa, and soybeans.
- Fuel is needed to manufacture herbicides and fertilizers.
- Gasoline is needed to plant seeds, spread fertilizer, and spray herbicides.
- Gasoline is needed to harvest crops and ship them to collecting points.
- Fuel is needed to distribute grains to farmers who are raising animals.
- Electricity is used to pump water from underground for the purpose of irrigating crops.
- Electricity is needed to process and refrigerate meat.
- Fuel is used to ship meat to stores, then to homes.

- Gas or electricity is used to cook meat.
- Fuel is used to cut down forests to make space for growing feed crops and ranching cows.
- Carbon dioxide is released from wood in trees when they burn or decompose (deforestation).
- Fuel is used to transport grains and animals across oceans to other countries.
- Carbon dioxide is released when cows digest their food.
- Gases are produced during the decomposition of animal wastes.

Which of these steps in the process of bringing meat to our plates should be considered in calculating the importance of livestock production in climate change? It is apparent that this is a complex issue, which can readily lead to disagreement in the scientific community. The FAO did not include all of the above criteria, hence its low number for the percentage contribution to climate change from livestock production. The Worldwatch Institute figure included all of these factors in coming to their conclusion of a 51 percent contribution from producing animal foods. But there is no doubt that all these activities are part of animal food production and might go into the calculation.

As an aside, food animals are raised under brutal conditions with very limited space, little or no medical care, poor food, beatings, non-anesthetic surgeries, and very short lives. Unfortunately, this is a universal approach to producing animals for food by large-scale, mass-production methods, commonly known as factory farming. These mass-production methods for producing animal foods require immensely more energy, farmland, water, fertilizer, herbicide, gasoline and waste removal than growing potatoes, peas and peaches. This is true for all of the plant foods that we commonly eat.[38]

In the United States alone, 10–11 billion chickens are raised for food. That comes to an average of about thirty chickens per person per year.

Worldwide pig production, reaching almost 1 billion animals each year, releases 668 million tons of greenhouse gases, and the statistic for chickens is 606 million tons.[39] Cow production for beef and milk is, by far, the greatest contributor to greenhouse gases. Keep in mind the total greenhouse gas production figure of 46 billion tons per year worldwide. Let's do a little calculation: for beef, just under thirty pounds of greenhouse gas are produced for each pound of meat consumed. Twelve million tons of beef are produced and consumed in the United States each year. Multiplying those numbers yields a total of approximately 350 million tons of greenhouse gases from beef consumption alone in the United States. A similar calculation for pork yields a figure of about 150 million tons of greenhouse gases.[40, 41] Simply put, enormous and dangerous amounts of carbon dioxide and methane are produced in fueling our meat consumption habits.

The FAO found, surprisingly, that the greenhouse gas emissions from livestock production are greater than from all forms of transportation combined, including cars, trucks, planes, and others. A switch away from meat-eating saves more greenhouse gases than a switch from an SUV to a hybrid vehicle.[42]

IV. Where Are the Media??

In the introduction, I asked why we haven't heard much about climate change. In this chapter, I will provide some ideas as to why this is happening. And it is extremely unfortunate, for our lack of knowledge plays a large part in making the climate change problem so much more difficult to solve.

There are some possible explanations as to why we in the United States have seen and read very little about these problems in the major media. Other countries are likely to have different situations.

* National polls show that half the United States population believes that climate change is of little or no harm while 43 percent deny that it is caused by humans as a result of burning fossil fuels in the enormous way that we've been doing the last two hundred years.[43] The media are reluctant to provide information unacceptable to a large portion of the public.
* Many of our representatives in Congress think climate change isn't scientifically established; some think it's a hoax, and some think there are other causes for climate change. They, too, are not likely to share information that so many people find objectionable.
* Our presidents haven't yet bought into climate change as an emergency requiring major national attention. Thus, we aren't informed about climate change from that source.
* Major media corporate officers may believe that the subjects are not yet politically correct.
* Major environmental groups such as Greenpeace and Sierra Club, although they do some excellent work, have not yet adopted climate change and livestock production as the most important issues of our time. A recent documentary movie, *Cowspiracy*, showed how they refuse to acknowledge the livestock industries as major contributors to greenhouse gas production.
* Television channels are owned by large corporations that may fear loss of audience and advertising revenue if they make presentations about climate change.
* Media corporations, in general, are likely to see attempts to solve the climate change problem as potential threats to the economy as well as to their income.
* Meat and dairy corporations would rapidly lose income if millions of people stopped consuming their products. It's not in their interest to tell you about the environmental

harm resulting from that consumption. These corporations have many connections with the media and with Congress.
* Likewise, corn and soybean farmers who supply the food for cows, pigs, chickens, and turkeys would be financially threatened if word got out about their contribution to climate change. Their influence in Congress is very powerful also.
* Changing our eating habits from eating animal flesh to mainly eating plant foods will make us a healthier nation, less needing of medical services. Pharmaceutical companies, hospitals, and doctors would face reduced incomes.
* Because meat and dairy consumption is so deeply a part of the eating pattern of people in industrialized nations, it is generally thought that people would not accept the connection between meat consumption and climate change, even if told by reliable sources. It is commonly understood that people would not be willing or able to make the change in eating that is necessary to stave off the threats of climate change. So why bother sharing that information if it's going to fall on deaf ears? The same is true in other industrialized nations.

* Psychologically, it is very difficult for individuals and organizations to suggest lifestyle changes that are seen as having long-term benefits. We tend to function best dealing with short-term changes.[44, 45, 46, 47, 48,]

Not a pretty picture, considering the seriousness of the situation in which we find ourselves, but these are the factors that tend to keep us ill-informed of what is happening to the world around us.

V. Conclusion

Allow me to address you with a personal message. If you have gone through these very serious chapters with their collection of ideas, information, and numbers, I would think that you would be shocked at the situation that faces all of us. We are looking at record numbers of storms with devastating effects, such as tornados, hurricanes, and typhoons; extreme droughts; millions of people dying from starvation, dehydration, and drowning; extreme loss of ocean life; perhaps billions of people losing homes and finding no place to go that is safe; mass extinctions of species and the inability of plant and animal life to adapt to the unstoppable heating of the air we breathe and the water that we need to drink. Certainly you now realize, if you didn't before you picked up this book, that life on our only planet is threatened beyond anyone's imagination. Land animals, airborne animals, ocean animals, and all kinds of plants are under extreme stress. And that includes every one of us.

Many governments, for-profit organizations, nonprofit organizations, and corporations have proposed various kinds of solutions: taxing carbon, constructing massive solar and wind turbine systems, and developing other sources of energy besides oil, coal, and natural gas. All of these proposals are fine and would be helpful, but there are built-in limitations that make them unlikely to solve the climate crisis! The climate emergency! The climate chaos! They require international government and corporate cooperation. They require vast amounts of money. They take decades to install. There are geographic and physical problems resisting accomplishment. All these detriments make it very unlikely that they will be in place soon enough and in enough locations around the world to halt or even slow climate change.

As you have now read, the livestock industries and the industries supporting them (corn, soybean, barley, and grass farming; fertilizer manufacturers; tractor, train, and truck manufacturers; herbicide producers, etc.) are making enormous, harmful changes in the quality of our atmosphere, dumping billions of tons of undesirable gases into it. As of this writing, there is no significant decrease in these industrial outputs. More and more countries want to have the standard of living that we in industrialized nations have. That means more and more carbon dioxide and methane. That means more and more heating of the air and water that every living thing requires for its existence. And yet, life goes on as if there is no threat to all living things on our planet! It is beyond comprehension.

Emergency situations require emergency responses. If your house is on fire, trying to put it out with pails of water won't work. If someone has a heart attack, you know that taking aspirin will not be sufficient. If someone invades your home with a gun, threatening that person with a stick will not be much of a deterrent. Our planet is in an emergency situation. Emergency responses are necessary!

In spite of all the terrible things happening around the world because of climate change, *there is one window of opportunity!* And only one! If we end our consumption of animal foods—the cows, pigs, chickens, and turkeys (in many other countries, you can add rabbits, goats, sheep, ox, and camels)—then there is a great chance that we can fend off the rising temperatures of our air and water. We must switch from meat-based eating to mainly plant-based eating. It is that simple. Yet it is so difficult due to the resistance of people to changing their eating patterns. But it must be done, now, today; at the latest, tomorrow. We need to reach a tipping point in adopting this eating style, where it becomes the normal way of eating, replacing today's meat-eating as the standard. So, whether livestock contributes 14 percent or 51 percent to the warming doesn't matter. The impact is large enough that ending animal agriculture could very likely save life on our planet: less likelihood of extreme hunger and dehydration, less death in the oceans and land, less flooding and displacement of people. These horrible events will take place, guaranteed, if we do nothing but watch.

Additional, very strong reasons for making the shift away from animal food consumption are that it can be done overnight, requires no government legislation or approval, entails no financial investment, and can be done by everyone, everywhere. And there are some fringe benefits thrown in: better health for all of us and our families and fewer animals tortured. People eating mostly plant foods have much less chance of developing chronic diseases such as heart disease, diabetes, some cancers, osteoporosis, obesity, and so many more ailments.

There are millions of people around the world who have changed their eating habits in this way. What do we eat? Lots of fruits, vegetables, whole grains, beans, nuts, and seeds. I have thousands of Facebook friends in the United States as well as from outside the United States who are also working on urging people to make the shift to plant-based eating. Many of these committed friends reside in Australia, Canada, Brazil, Germany, France, Italy, Portugal, Spain, England, Scotland, Ireland, Denmark, Norway, Sweden, Belgium, Hungary, Turkey, Syria, Israel, India, Pakistan, South Africa, Ghana, China, Japan, Malaysia, Indonesia, and New Zealand. Now hundreds of millions more people are needed to make the change. With all of you reading this book, I hope that we will come closer to achieving the goal of saving all life on the planet.

I have attempted to show you why this change is absolutely

necessary for survival and how it can be accomplished easily by each of us. Future generations will thank us.

Bibliography

Bowen, Mark. *Thin Ice, Unlocking the Secrets of Climate in the World's Highest Mountains.* New York: Henry Holt & Co., 2005.

Diamond, Jared. *Collapse, How Societies Choose to Fail or Succeed.* New York: Penguin Group, 2005.

Flannery, Tim. *The Weather Makers; How Man Is Changing the Climate and What It Means for Life on Earth.* New York: Atlantic Monthly Press, 2005.

Glavin, Terry. *The Sixth Extinction, Journeys Among the Lost and Left Behind.* New York: Thomas Dunne Books, 2006.

Hansen, James. *Storms of My Grandchildren, the Truth About the Coming Climate Catastrophe and Our Last Chance to Save Humanity.* New York: Bloomsbury, 2009.

Klare, Michael. *The Race for What's Left, The Global Scramble for the World's Last Resources.* New York: Henry Holt & Company, 2012.

Lappe, Anna. *Diet for a Hot Planet, the Climate Crisis at the End of Your Fork and What You Can Do About It.* New York: Bloomsbury, 2010.

Lynas, Mark. *Six Degrees, Our Future on a Hotter Planet.* Washington: National Geographic, 2008.

Oreskes, Naomi, and Erik M. Conway. *Merchants of Doubt, How a Handful of Scientists Obscured the Truth on Issues from Tobacco Smoke to Global Warming.* New York: Bloomsbury, 2010.

Singer, Peter, and Jim Mason. *The Way We Eat, Why Our Food Choices Matter.* Rodale, 2006

Weart, Spencer R. *The Discovery of Global Warming.* Cambridge: Harvard University Press, 2003.

Endnotes

1. Moses Seenarine, *Meat Climate Change,* unpublished as of May 2015.
2. "L.A.'s Air is Getting Dirtier: The Drought Impact You Haven't Heard About," *Los Angeles Times*, Jan. 21, 2014, http://www.weather.com/travel/commuter-conditions/news/los-angeles-dirty-air-20140121.
3. Mark Lynas, *Six Degrees: Our Future on a Hotter Planet* (Washington, D.C., National Geographic Society, 2008).
4. Robert W. Felix, "Abrupt Climate Change for Past 800,000 Years.," Sept. 11, 2011, accessed May 13, 2015, http://iceagenow.info/2011/09/abrupt-climate-change-800000-years/.
5. "Major Storms of the Past 10 Years," Nov. 8, 2013, CNN, accessed May 13, 2015, http://www.cnn.com/2013/11/08/world/gallery/storm-radars/index.html.
6. "Is Global Warming Linked to Severe Storms?" Union of Concerned Scientists, accessed April 28, 2015, http://www.ucsusa.org/global_warming/science_and_impacts/impacts/global-warming-rain-snow-tornadoes.html#.VVOQnvlViko.
7. David Adam, "Amazon could shrink by 85% due to climate change, scientists say," March 11, 2009, *Guardian*, accessed May 13, 2015, http://www.theguardian.com/environment/2009/mar/11/amazon-global-warming-trees.
8. "Deserts spreading across globe, U.N. warns," 6/16/2004, NBC News, accessed May 14, 2015, http://www.nbcnews.com/id/5226259/ns/us_news-environment/t/

deserts-spreading-across-globe-un-warns/#.VVS4NPlViko.
9. Ker Than, "Animals and Plants Adapting to Climate Change," June 20, 2005, Live Science, accessed May 14, 2015, http://www.livescience.com/3863-animals-plants-adapting-climate-change.html.
10. Tim Flannery, *The Weather Makers; How Man Is Changing the Climate and What It Means for Life on Earth* (New York: Atlantic Monthly Press, 2005).
11. Robert H. Westover, "Impact of Climate Change on Forest Diseases Assessed in New US Forest Service Report," May 18, 2012, US Department of Agriculture, accessed May 13, 2015, http://blogs.usda.gov/2012/05/18/impact-of-climate-change-on-forest-diseases-assessed-in-new-us-forest-service-report/.
12. Joe Romm, "Science: Ocean Acidifying So Fast It Threatens Humanity's Ability to Feed Itself," March 2, 2012, Climate Progress, accessed May 13, 2015, http://thinkprogress.org/climate/2012/03/02/436193/science-ocean-acidifying-so-fast-it-threatens-humanity-ability-to-feed-itself/.
13. "Great Barrier Reef Animals and Marine Life," accessed May 13, 2015, http://www.greatbarrierreef.com.au/animals.
14. "Ocean Heat," May 2014, Environmental Protection Agency (EPA), accessed May 13, 2015, http://www.epa.gov/climatechange/pdfs/print_ocean-heat-2014.pdf.
15. "Warmer Oceans," Aug. 8, 2014, a student's guide to Global Climate Change, accessed May 16, 2015, http://www.epa.gov/climatechange/kids/impacts/signs/oceans.html.
16. "Sea Level Rise," National Geographic, accessed May 16, 2015, http://ocean.nationalgeographic.com/ocean/critical-issues-sea-level-rise/Environmental Protection.
17. Daniel DeNoon, "Salt-Water Fish Extinction Seen by 2048," Nov. 6, 2006, CBS News, accessed

May 16, 2015, http://www.cbsnews.com/news/salt-water-fish-extinction-seen-by-2048/.
18. "The Extinction Crisis," Center for Biological Diversity, accessed May 16, 2015, http://www.biologicaldiversity.org/programs/biodiversity/elements_of_biodiversity/extinction_crisis/.
19. John Abraham and Jeff Masters, "Climatic tipping points, stories about our possible future," Oct. 19, 2013, *Guardian*, accessed May 16, 2015, http://www.theguardian.com/environment/climate-consensus-97-per-cent/2013/oct/19/climate-change-tipping-points-stories-future.
20. Dr. James Hansen, "Twenty Years Later: Tipping Points Near on Global Warming," July 1, 2008, *Huffington Post*, accessed May 16, 2015, http://www.huffingtonpost.com/dr-james-hansen/twenty-years-later-tippin_b_108766.html.
21. Caitlyn Kennedy, "2012 State of the Climate: Sea Surface Temperatures," July 30, 2013, NOAA Climate.gov, accessed May 16, 2015.
22. "Carbon Dioxide," NASA Global Climate Change, accessed May 16, 2015, http://climate.nasa.gov/vital-signs/carbon-dioxide/.
23. "Atmospheric methane," Wikipedia, accessed May 16, 2015, http://en.wikipedia.org/wiki/Carbon_dioxide_in_Earth%27s_atmosphere.
24. Ibid.
25. Steve Hamburg, "Methane: the other important greenhouse gas," Environmental Defense Fund, accessed May 16, 2015, http://www.edf.org/climate/methane.
26. "Methane and Carbon Dioxide Global Warming Potential," accessed May 16, 2015, http://www.global-warming-forecasts.com/methane-carbon-dioxide.phpf.
27. "Global Greenhouse Gas Emissions," EPA, accessed May 16, 2015, http://www.epa.gov/climatechange/science/indicators/ghg/global-ghg-emissions.html.

28. "Future Climate Change," EPA, http://www.epa.gov/climatechange/science/future.html.
29. "Some Predicted Future Effects of Climate Change," Climate Concern UK, accessed May 16, 2015, http://www.climate-concern.com/Predicted%20Future%20Effects.htm.
30. Brandon Keim, "Using Climate Change to Predict Wars," May 31, 2007, *Wired,* accessed May 16, 2015, http://www.wired.com/2007/05/using_climate_c/.
31. "Predictions of the human cost of climate change," Feb. 8, 2013, *Science News,* accessed May 16, 2015, http://www.sciencedaily.com/releases/2013/02/130208105309.htm.
32. "Food Choices and the Planet," Earth Save: Healthy People Healthy Planet, accessed May 16, 2015, http://www.earthsave.org/environment.htm.
33. "Importance of Livestock Agriculture in the US," US Dept. of Agriculture (USDA), accessed May 16, 2015, http://www.training.fema.gov/emiweb/downloads/is111_unit%202.pdf.
34. "Major cuts of greenhouse gas emissions from livestock within reach," Sept. 26, 2013, Food and Agriculture Organization of the United Nations (FAO), accessed May 16, 2015, http://www.fao.org/news/story/en/item/197608/icode/.
35. "Livestock a major threat to environment," Nov. 29, 2006, FAO, accessed May 16, 2015, http://www.fao.org/newsroom/en/news/2006/1000448/index.html.
36. Robert Goodland and Jeff Anhang, "Livestock and Climate Change," November/December 2009, Worldwatch Institute, accessed May 16, 2015, http://www.worldwatch.org/files/pdf/Livestock%20and%20Climate%20Change.pdf.
37. "How Factory Farms Impact You," Food & Water Watch, accessed May 16, 2015, http://www.factoryfarmmap.org/problems/.

38. "Greenhouse gas emissions from pig and chicken supply chains," FAO, accessed May 16, 2015, http://www.fao.org/docrep/018/i3460e/i3460e.pdf.
39. "Climate and Environmental Impacts," Environmental Working Group, accessed May 16, 2015, http://www.factoryfarmmap.org/problems/.
40. "Cattle & Beef," United States Department of Agriculture Economic Research Service, accessed May 16, 2015, http://www.ers.usda.gov/topics/animal-products/cattle-beef/statistics-information.aspx.
41. Damian Carrington, "Eating less meat essential to curb climate change, says report," Dec. 2, 2014, *Guardian,* accessed May 16, 2015, http://www.theguardian.com/environment/2014/dec/03/eating-less-meat-curb-climate-change.
42. "Climate Change Belief," May 19, 2015, *Huffington Post,* accessed May 19, 2015, http://www.huffingtonpost.com/news/climate-change-belief/.
43. Phil Plait, "The Utter Predictability of Climate Change Denial," May 8, 2014, *Slate,* accessed May 16, 2015, http://www.slate.com/blogs/bad_astronomy/2014/05/08/climate_change_denial_after_global_warming_report_deniers_deny.html.
44. "Climate Change Overview," March 24, 2015, World Bank, http://www.worldbank.org/en/topic/climatechange/overview.
45. Ross Koningstein and David Fork, "What It Would Really Take to Reverse Climate Change," Nov. 18, 2014, IEEE Spectrum, accessed May 16, 2015, http://spectrum.ieee.org/energy/renewables/what-it-would-really-take-to-reverse-climate-change.
46. Travis McKnight, "Want to have a real impact on climate change? Then become a vegetarian," August 4, 2014, *Guardian,* accessed May 16, 2015, http://www.theguardian.com/commentisfree/2014/aug/04/climate-change-impact-vegetarian.

47. Brandon Keim, "The Psychology of Climate Change Denial," Dec. 9, 2009, *Wired,* accessed May 16, 2015, http://www.wired.com/2009/12/climate-psychology/.
48. "Is denial causing fewer people to become vegetarian?" Feb. 16, 2010, Food, Farm, and Famine, accessed May 16, 2015, http://rpi-fff.blogspot.com/2010/02/is-denial-causing-fewer-people-to.html.

 Len Frenkel is a retired chemist and high school science teacher and lives with his wife in Bethlehem, PA. He has been an environmentalist for at least forty-five years, concerned about climate change for at least twenty years, and a vegetarian for twenty-four years. He recently edited a textbook on the subject of climate change and has written many notes and letters to the editor on this subject. This is his first, and probably last, book.

www.ingramcontent.com/pod-product-compliance
Lightning Source LLC
Chambersburg PA
CBHW041132200526
45172CB00018B/149